Zoom for Beginners (2020 and Beyond)

A No-Fluff Guide for Teachers and Businesses to Master the Use of Zoom for Virtual Meetings, Online Classes, Webinars and Video Conferencing

By

Chris Button

Copyright © 2020 – Chris Button
All rights reserved

No part of this publication may be reproduced, distributed, or transmitted in any form or by any means, including photocopying, recording, or other electronic or mechanical methods, without the prior written permission of the publisher, except in the case of brief quotations embodied in reviews and certain other non-commercial uses permitted by copyright law.

Disclaimer

This publication is designed to provide competent and reliable information regarding the subject matter covered. However, the views expressed in this publication are those of the author, and should not be taken as expert instruction or official advice from Google. The reader is responsible for his or her own actions.

The author hereby disclaims any responsibility or liability whatsoever that is incurred from the use or application of the contents of this publication by the purchaser or reader.

Books By The Same Author

Google Classroom for Teachers (2020 and Beyond)

Table of Contents

Books By The Same Author 2

Introduction 7

Chapter 1 10

ABC of Zoom 10

    What is Zoom? 10
    Who Can Use Zoom? 10
    Main Features of Zoom 11
    Devices Supported By Zoom 12
    Pros and Cons of Zoom 13

Chapter 2 16

Zoom Account Types and Plans 16

    Zoom Paid Users 16

        Zoom Pro 16
        Zoom Business 17
        Zoom Enterprise 17
    Zoom Free Users 19

    Zoom Plans and Pricing 19

        Zoom for Education 19

Zoom for Telehealth 20
Zoom for Developers... 20
Chapter 3 21
Getting Started With Zoom 21
Downloading and Installing Zoom App 21
For Desktops and Laptops 21
For Android Devices 21
For iOS (iPhone and iPad)... 22
Signing Up For a Zoom Account........ **Error! Bookmark not defined.**
Scheduling a Zoom Class or Meeting 30
Starting a Zoom Class or Meeting 33
Inviting Students/Participants to a Zoom Meeting 34
Via email 34
Via contacts 35
Via invite link 36
Preventing Zoombombing 37
Recording a Meeting 44
Sharing Your Screen 49
On Windows and Mac... 50
On Android and iOS.. 53
Using Zoom Whiteboard 55

What You Need for Zoom Whiteboard 56
Accessing the Whiteboard Feature................. 56
Saving a Whiteboard Session........................... 58
Using Zoom Annotation 59

How to Enable Annotation 59
Annotation Tools 61
Setting Up Breakout Rooms.............................. 63

How to Enable Breakout Rooms 63
How to Create Breakout Rooms....................... 65
Managing Breakout Rooms 65
Setting Up and Using Polling For Student/Participants... 69

Setting Up and Using Quizzes For Students/ Participants 75

Using Zoom Chat Tool.. 79

With Your PC 79
With Your Mobile Device.................................. 79
Activating Verbal and Non-Verbal Feedback 80

Transcribing Zoom Recording.......................... 84

Ending a Meeting 87

Scheduling and Using Zoom Webinar............ 87

Best Practices While in Zoom Class or Meeting................ 90

Chapter 4 93

Troubleshooting Most Common Zoom Problems 93

Chapter 5 99

Cool Tips and Tricks to Enhance Productivity With Zoom. 99

Chapter 6 111

Zoom Frequently Asked Questions (FAQs) 111

Conclusion 115

Introduction

Some time ago, several tools that facilitated distance communication and teleconferencing began hitting the consumer market. For instance, the first phone that was able to transmit video signals throughout the U.S was created in 1964. This tool was quite expensive and only the well-to-do class could afford such. Although this was a great achievement in technology, it still had its limitations.

Fast track to today's world, many tools have been developed to bridging the communication divide and limitations encountered in the earlier years. Within a split second, actually, in milliseconds, you can connect with someone else thousands of miles across the globe. With today's modern technology, coupled with the internet, distance communication in real-time has been made so effortless and seamless, just as your breathing. Asides this, communication no longer has to be restricted to the elites nor has to be expensive as it usually was. With video conferencing made possible, communication with 2 or more persons irrespective of their location has been made easily affordable and accessible without having to pay through their nostrils.

Special applause to technology that brought about the innovation of these tools. One such tool that has changed how we communicate and provided a seamless way to carry out our everyday physical activites but in the virtual space is Zoom.

Zoom is a teleconferencing tool that allows you to engage in real-time, peer-peer communication over the internet, with the option to choose either to meet with people via video or audio. The Zoom app allows communication between persons to instantly take place even if they are thousands or even millions of miles away from each other.

From conducting distance education to organizing business events, having corporate meetings, and engaging in social relations, Zoom provides its users some form of flexibility to effortlessly interact with each other without having to go through several hassles.

This book takes a deep dive into the inner workings of the Zoom tool, especially as it relates to not only driving online education among teachers and students but also amongst business partners and co-workers alike. This book covers the basics of what you need to know about the Zoom application and some pretty cool advanced stuff to get you by with making the most of this tool more effectively, anytime, any day.

At the end of this guide, you will;

- Have mastered the Zoom tool to make the most of it for your virtual meetings, classes and online events.
- Know how to use Zoom for audio and video conferencing and for hosting webinars.
- Be enlightened on the Zoom account types and plans you can subscribe to for the best virtual and online meeting experience.
- Know how to set up and make use of Zoom effectively to ensure your virtual meetings and classes are as successful, fun, and engaging as they can be.

And much more!

There is just a whole lot of things you can do with Zoom, which you probably have no idea of. Well, you are about to find out every one of them right about now. So, grab your chair, sit back, and relax while I take you on a roller-coaster ride on how to max out the features of Zoom to provide the best virtual and online meeting experience you can possibly think of.

Chapter 1

ABC of Zoom

What is Zoom?

Zoom is a video and audio conferencing application that enables virtual interaction between two or more people in an educational or business setting. It gained prominence during the Covid-19 pandemic-induced lockdown, as it created an avenue for organized events, co-worker meetings, conferences and many others to hold virtually.

There were even cases of Zoom parties! Depending on your Zoom package, it is possible to host a meeting with more than 300 people.

Who Can Use Zoom?

There is no restriction to using Zoom; as long as you have a device suitable for the application, then you can sign up to the platform. The process of setting up your virtual meeting is easy enough; you don't need to be tech-savvy to do so.

Large, medium and small-scale businesses can use Zoom to hold meetings with their workers and teachers alike can use it for virtual classes with their students. Seminars and conferences can be held on Zoom with participants' attending' from all over the world.

Main Features of Zoom

One of the main features of Zoom is its HD (High Definition) video and audio feature. This gives the app its clear video and audio quality no matter the package you subscribe to.

Another feature is the full screen and multi-screen view; it allows you to view only one person or many people at once, depending on your preference.

The HyperText Transfer Protocol (HTTP) access, Advanced Encryption Standard (AES) 256-bits encryption, Secure Sockets Layer (SSL) encryption are features that ensure the security of all your data. It also allows for webinars, in-meeting chat and video or audio calls.

Depending on your payment package, Zoom meetings or rooms can hold up to 1,000 participants at once. Another feature is its local and Cloud-based video and audio recording with transcripts.

Some other features are the Zoom whiteboard (for highlighting and annotation), Zoom breakout rooms (to divide Zoom meeting participants into smaller and more manageable classes), the polling feature (this allows you to organize polls among participants). There are also the verbal and nonverbal features that will enable you to receive feedback from meeting participants.

Devices Supported By Zoom

Zoom supports various devices depending on the operating system of the device.

Apple Devices
- iPhone 4 or higher
- iPad Pro
- iPad Mini on iOS 8.0 or higher
- iPad 2 or higher
- iPod touch 4th Gen
- iPhone 3GS
- iPad OS 13 or higher

For Android, devices with Android 5.0x or higher will do for Zoom

Computer devices with the following operating system will support Zoom:

- Windows 10 Home, Pro or Enterprise
- Windows 8 or 8.1
- Windows 7
- Ubuntu 12.04 or higher
- Mut 17.1 or higher
- Red Hat Enterprise Linux 6.4 or higher
- Oracle Linux 6.4 or higher
- Open SUSE 13.2 or higher
- ArchLinux (64-bit only)

Pros and Cons of Zoom

Like everything else, there are advantages and disadvantages to using Zoom. Some of the advantages of using Zoom are:

- High-quality audio and video output
- It's easy to navigate its interface
- Supports long video calls and streamlined video conferencing
- Top-notch security feature

- It is Health Insurance Portability and Accountability Act (HIPAA) compliant
- It allows for the integration of partner software
- It is expandable to allow for a large number of participants
- It allows you to share multifunctional screen
- You can sync the app with your calendar
- It has a virtual background feature
- It works with various operating systems like Apple, Windows and Android
- It allows you to share slides and other content

Zoom's major disadvantage is that its video conferencing service is quite heavy and can greatly overwork a user's computer or mobile device's CPU (Central Processing Unit). Also, if you don't have a good internet connection, you can't take full advantage of the Zoom features.

Chapter 2

Zoom Account Types and Plans

Zoom Paid Users

There are three different subscription packages for paid users: Pro, Business and Enterprise.

Zoom Pro

This package is for small businesses or teams; it costs $149.90 per year for each license. With Zoom Pro, you can host up to a hundred participants at once, have unlimited group meetings and you can live stream your Zoom meetings to almost every social media platform.

This package also offers 1 gigabyte Cloud recording for every license you purchase and with your Zoom account, you can purchase up to nine licenses. It also has more user management and admin controls compared to the Zoom basic package.

System administrators on the Zoom Pro license have access to the report's sections so they can see how many meetings the business or organization is having, how many participants are attending the meetings and the duration of the meeting.

Some other features available on Zoom Pro are SIP/H.323 room connectors, Zoom Rooms, and Zoom webinars.

Zoom Business

This package is for small and medium-sized business or teams. It costs more than the Pro package at $ 199.90 per year for each license.

On Zoom business, you can host up to three hundred participants and with this, comes with a single sign-on and Cloud recording transcripts feature.

Other features that come with the Business package are managed domains that allow for company branding.

Zoom Enterprise

Zoom Enterprise is meant for large businesses and teams. It costs the same as the Business package ($199.90 per year for a license), but it can do a whole lot more than the business package.

For one, you can host up to five hundred participants at once and about a thousand participants if they are on the Enterprise package.

It offers unlimited Cloud storage, a dedicated customer success manager and a transcription package.

There are optional add-on plans you can also subscribe to, but you have to have at least one licensed user to purchase these add-ons. They are:

1. Audio Plans

This costs $100 per month; it allows you add CallOut, global Toll-free and local dial-in for premium countries.

There's no charge to any of your participants to call in from any device, and it also allows you to select single or multiple countries for toll-free call-in easily.

2. Cloud Storage

This additional plan allows you store, stream and download all your videos from Zoom Cloud. Different video file formats like MP4 and M4A are available.

The Cloud storage add-on gives storage options of up to three terabytes per month and it costs $480 per year.

3. Large Meeting

This package starts at $600 per year; it increases the participants' capacity of your meeting plans. For example, if you are on a Pro plan, the Large Meeting add-on will increase the capacity from a hundred to five hundred.

It allows up to a thousand interactive participants per meeting and a flexible payment plan with monthly or yearly billing options.

Zoom Free Users

The package for free users called Zoom Basic is completely free with no hidden charges or trial periods.

You can host up to a hundred participants in a maximum of forty-minute group meetings. It also offers unlimited one on one meetings.

Zoom Plans and Pricing

There are three main Zoom plans: Zoom for Education, Zoom for Telehealth and Zoom for Developers.

Zoom for Education

This plan is for schools at all levels. In line with the global pandemic and its attendant effects, Zoom decided to remove the 40-minute time limit on all Basic accounts for primary and secondary schools on the Zoom for Education Plan.

The plan also offers secure video communication for e-learning that aids improved learning from nursery to university level.

Zoom has flexible Education plans that start from $1,800 annually for twenty hosts. They also have resources that aid education like the higher education datasheet, the K-12 datasheet and Federal Educational Rights and Privacy Act (FERPA) guide.

Zoom for Telehealth

This plan allows for video conferencing that connects you to compliant Health Insurance Portability and Accountability Act (HIPAA)/Personal Information Protection and Electronic Documents Act (PIPEDA) plans. It begins at $200 per month per account with 10 hosts. There are also 1, 2 and 3 year prepaid packages.

Zoom for Developers

This comes in the form of Zoom App Marketplace, an open platform for third-party developers that allows them to build integrations and applications on the spine of Zoom's video-first unified communications platform.

It also allows them to take advantage of Application Programming Interfaces (APIs), Software Development Kits (SDKs) and Webhooks to build custom-made applications that can boost your business with a wide range of collaboration suite.

The plan starts at $100 for four thousand minutes of use.

Chapter 3

Getting Started With Zoom

Downloading and Installing Zoom App

For Desktops and Laptops
1. Open your browser and go to http://Zoom.us
2. Go to "Download Center and click the download button under "Zoom Client for Meeting."
3. Automatically, the app will download on your device.
4. After downloading, follow the prompts to install the application on your device and you are ready for your first meeting.

For Android Devices
1. Go to the Google Play Store App on your device or open your browser and type in play.google.com
2. When the app or page opens, tap on the "Apps" button or icon. On the top of the screen, there's a

search taskbar (you recognize it by the magnifying glass at the right corner)
3. Type in "Zoom Cloud Meeting" then tap on search.
4. When the results are shown, tap on the appropriate one.
5. Tap on install.
6. On the next page, tap Accept.
7. The application will automatically download and install on your device.
8. When it is installed, tap the open button.
9. When the app opens, tap on the sign-up button and fill in your details.
10. Follow through with the on-screen instructions to complete the registration process and sign in

For iOS (iPhone and iPad)
1. Open the App Store and search for "Zoom Cloud Meeting."
2. Tap on the right option from the search results.
3. Tap install and the app will automatically download and install.
4. When it is done, tap on open

5. When the app opens up, tap the sign-up button and fill in your email address and password, and follow through with the on-screen instructions to complete the registration process and sign in.

Signing Up For a Zoom Account

Signing up for a Zoom account is an easy task. You can get this done using any device that supports the Zoom app, including a mobile device, Windows or Mac device.

Using Your Mobile Device

If you are using a mobile device (iOS or Android devices), the steps are similar across these platforms.

1. After downloading the Zoom app as described in the section above, launch it. On the welcome page, there are two options "sign up" or "sign in." Click on sign in.

2. On the new page displayed, type the email address you prefer using for the Zoom account. Ensure it is a valid one because a confirmation mail will be sent to your email address to complete the registration process.

3. Afterward, tick off the tiny box that says, "I agree to the terms of service." When you are done, click the "sign up" button. A notification will pop up notifying you that a confirmation email has been sent to your email address. Tap "ok" to close this notification.

4. Check your email and click the "activate account" button. When this is done, the activation link will be opened in your phone's browser.

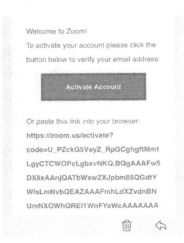

At this point, commence the confirmation stage by ensuring all details entered are valid. After doing this, choose a password that is secured enough.

Note: Your password must contain the following:
- Should be 8 characters
- At least one number
- At least one letter
- Both upper and lower cases

6. When you are done with the above step, proceed to invite your colleagues or students. The next step, after successfully setting your password encourages you to begin a meeting immediately. If you don't want to do this just yet, you can skip this step.

7. The step after this also encourages you to begin a meeting immediately. Click on "go to my account" if this is still not what you want to do - this will take you to where you will type in the details used to sign up for the Zoom service; your email address and password. When done with, click on "Sign In" to gain access to the account you just created and get started with using the Zoom app.

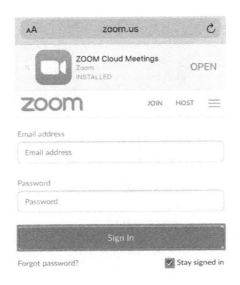

Using Your Windows Browser

Follow the steps outlined below to sign up for Zoom using your Windows browser.

To sign up for a Zoom account, take the following steps:

1. Type in the URL http://Zoom.us on your browser
2. When you get to the webpage, click on the Sign-Up button
3. You can register with your email address, your Facebook account or your Google account.
4. If the Zoom account is meant to represent your business, it is preferable to open with your

business email address if you have one. Make sure the email address presented is an active one as an activation email will be sent to the email address.
5. After clicking the button, you will be asked to input your date of birth; this is because people under 16 are ineligible to open a Zoom account. This is to protect young children from indecent interactions online.
6. When this is done, a mail is sent to the email address you registered with. The mail contains an "Activate Account" bar, click on the bar to activate your account. If you experience any issue with receiving the activation mail, check your spam folder. If the mail isn't in the spam folder, recheck the email address you registered with for correctness.
7. Suppose you receive the mail, but the activate account button is not responding. In that case, there is a link in the mail sent alongside the activate button, copy and paste this link into your browser – this will also activate your account.

8. When the webpage opens, you will be asked if you are signing up for a school. Click yes, if you are and no, if you are not.
9. When you have answered the questions you will be led to another page where you are requested to fill in your first and last names along with your password.
10. Your password must contain the following:
    - Should be 8 characters
    - At least one number
    - At least one letter
    - Both upper and lower cases

If you have trouble remembering your usernames and passwords, write it down and keep them safe.

11. A new page asks you to invite people to create their own Zoom accounts through email. This step is not necessary, so you can skip it.
12. After skipping, a URL is provided that takes you the page where you can start your own meeting. On that page, there is a "start meeting now" button.

13. Click on this button or copy the URL and paste in your browser. Either way, you will receive a prompt that asks you to download the Zoom desktop application.
14. Follow the steps to download and install the app.
15. When you are done with app installation, you will see the following options "Join a Meeting" and "Sign In". Since you want to start your own Zoom meeting, click on "Sign In"
16. You will be asked to provide your email address and password, the same ones you registered with.
17. If you signed in using Facebook or Google, there are also sign-in options for you.
18. When you have logged in successfully, go to "Home", and look for the orange "New Meeting" button and click on it.
19. You are ready to start your first Zoom meeting

Scheduling a Zoom Class or Meeting

It is very important to know how to schedule a Zoom meeting. It is like scheduling a physical meeting on your calendar, but this time, a virtual one.

Scheduling meetings in Zoom is not a difficult task. To get this done, all you need to do is follow the processes outlined below;

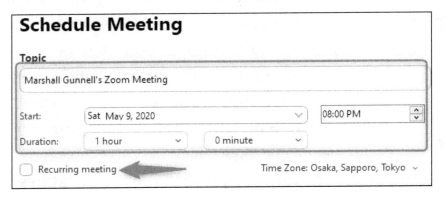

1. The first step in scheduling a Zoom meeting is to sign into the Zoom application. Once you are signed in, click the schedule button on the home page.
2. When the Schedule Meeting page opens, set the date, duration and name of meeting.
3. You can tick on Recurring Meeting at the bottom left corner of the window if the meeting occurs regularly. Be sure to notice and adjust the time zone at the bottom right of the window accordingly.

4. After doing the above, the next step is to get your Meeting ID. It is safe to automatically generate your ID to guard against Zoom bombing, so click on the "Generate Automatically" option.
5. For an added layer of security, you can add a password.
6. Then you adjust the video and audio settings to your preference
7. A prompt will ask if you would like to send the meeting invite and reminder to your Outlook, Google or any other calendar, click on your preferred option.
8. Next is the advanced options where you can enable or disable features like Waiting Room, Mute Participants On Entry, Adding A Co-Host, Join Before Host, Only Authenticated Users Can Join etc. Go through each option, enable, and disable as you wish.
9. When you are satisfied with everything, click on the Schedule button to schedule your Zoom meeting.

Starting a Zoom Class or Meeting

To start a Zoom class or meeting:

1. Sign in to your Zoom account and click on the "Meeting" tab on the home page.
2. On the meeting page, click on "Upcoming Meetings."
3. Click on the meeting or class you want to start, then click Start.
4. Once this is done, you will be directed to the meeting. Follow the on-screen instructions displayed to gain access to the class you want to access.

However, if the meeting you want to access is yet to be created, go through the steps listed in the section above, starting from signing up for the Zoom service, and then following through with the instructions earlier discussed.

Inviting Students/Participants to a Zoom Meeting

There are three methods to inviting participants to your Zoom class or meeting; via mail, sending invites through contact, and copying and sending the invite link through other means.

Via email

1. Click or tap on "Meetings" from the home page after signing in to the Zoom service
2. From the Meetings page, click on the "Participants" option
3. At the bottom of the panel that appears, click on "Invite"
4. Three options will appear, click on the email tab
5. Select your preferred email service provider.
6. A message is automatically composed in the body of the mail with the invite link
7. Select your recipients and send the mail.

A copy of the mail looks like this

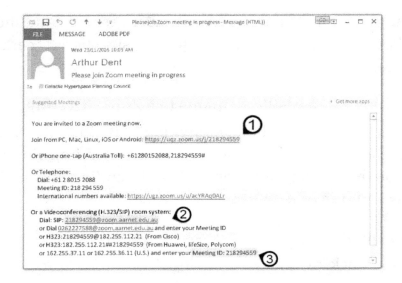

Via contacts

1. Click or tap on "Meetings" on the home page after signing in to Zoom service
2. Go to "Participants"
3. Click on "Invite"
4. Select the contact option
5. Search for and mark the contact or contacts you want to send the invite to
6. After selecting, click on "Invite," the selected contacts will receive a message with the invite link.

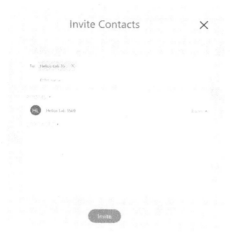

Via invite link

1. Click on "Meeting" on the home page after signing in
2. Go to "Participants"
3. Click on Invite
4. Highlight the invite link or the whole invitation text
5. Copy the highlighted link or text
6. Paste the link or text on whatever secure platform you want to send the invite to the intended participants.

Preventing Zoombombing

Zoombombing or Zoom raiding is gaining popularity and notoriety in virtual meeting space. It is the act of sending unwanted files, images, videos by mostly uninvited participants (who are usually internet trolls or hackers) to disrupt a meeting. It is like spamming a virtual meeting. In some cases, offending images like pornographic content, racially offensive images, or slurs, distracting sounds, or music are flashed during a Zoom meeting.

No matter how hard you try, it is likely that your Zoom meeting might end up a big mess. To curtail these Zoom bombers, follow the steps below:

1. Never use your Personal Meeting ID (PMI); your PMI is like your Zoom phone number. When sending an invite to a meeting, do not use your PMI because it makes it easier for the bomber to target you. There is an option to automatically generate a meeting ID. That way, for every meeting, or class you hold, a new meeting ID will be generated, making it hard for your meeting to be intruded by these Zoom bombers.

2. Always set a password; a password word is like two-step verification. After sending the automatically generated meeting ID, add a password to it. Make sure to send the password in a different message from the ID. That way, if the ID gets into the wrong hands, they will still need a password to attend your meeting.

3. Activate Zoom Waiting Room; the Waiting Room feature was introduced by Zoom in March 2020. As the name implies, it is a sort of reception area where you can screen participants or attendees. When the feature is enabled, the participants are in a virtual queue, waiting, until you are ready for them. This way, you can screen them individually before admitting them to the meeting.

   To set up the Waiting Room feature, go to "Advanced Options," from the options, click on the "In Meeting" icon and scroll until you find Waiting Room. Enable the feature. When the feature is enabled, you can customize it by choosing those that will be sent to the waiting room. You can also decide what message or image you want to display in your waiting room.

It might be your school logo and your school's mission statement, or the Zoom class outline, anything to engage the participants while they wait.

4. Mute audio and disable video for participants to prevent offensive or distracting sounds, unwanted videos, or participants' images. That way, the only image and sound participants can hear are those of the hosts and co-hosts. To achieve this, click or tap on the "Manage Participants" at the bottom of the meeting screen.

Other options will come up on the right side of your screen, scroll to the "More" option, and click on it click or tap on "Mute Participants on Entry." In addition, ensure the "Allow Participants To Unmute Themselves" option is disabled.

5. Turn off screen sharing for all participants except the host. Taking control of the screen is important to Zoombombers. Taking away that option is, therefore, paramount. Ensure that no one, except the host has access to the screen-sharing feature.

To do this, look for the Share Screen button at the buttom menu bar during the meeting. Click or tap on the arrow beside it to open the video menu, scroll to the Advanced Sharing Options.

When it opens, scroll to "Who Can Share" and set to "Only Host." Now only the host has access to the screen-sharing feature.

6. Update your app regularly. Zoom regularly updates its app, correcting some flaws in the older versions and adding new security and enhancement features in the newer versions. The company announced recently that they dedicated

all their engineering resources to fighting online bullying and harassment while also making their security features better so that users are better protected.

This why it is important to update your app, ensure you check your App store or Google store regularly to see if there is an update on the current app if you make use of the mobile version

7. Create an Invite-only Meeting; this feature is only available to subscribers on Zoom Pro and Enterprise. It is possible to restrict participants to a Zoom meeting by enabling the Invite-only feature.

    This ensures that the participant has to sign in using the email address you sent the invite to. To achieve this, go to options, then "Authentication Profiles," and enable it. When anyone who is not invited tries to join, they get a notification telling them that the meeting is only for invited attendees.

8. Lock Meeting once it starts. Once a meeting starts and everyone has joined the meeting, you can lock the meeting so that new participants cannot join the meeting. While the meeting is going on, go to the menu bar at the bottom of the screen and click or tap on "Participants," scroll through the options until you get to "More." More options will pop up, scroll to Lock Meeting and enable the feature.
9. Kick offenders out. Despite your best efforts, some offenders can slip through the cracks. Part of the features available to the host is the option to kick out an offender or, in a mild case, put the person on hold.

   To kick someone out, go to the menu bar, click or tap on participants and hover on the name of the person you want to remove. From the various options that appear, click or tap on "Remove."

   When someone is removed, the person can't rejoin. Some advice against allowing removed participants to rejoin the meeting, but in the case of wrongful removal, there is a feature that allows participants to rejoin when enabled.

This feature has to be activated before the meeting. To do this, from the home page, click or tap on the settings options. When the page opens, click or tap on the Meeting option, then enable In-Meeting (basic) or switch on the "Allow Removed Participants to rejoin."

If you think kicking out or removing participants is extreme, you can temporarily ban or put a participant on hold. To do this, look for the video thumbnail of the participant, click on it and select 'Start Attendee on Hold".

When you think they have learned their lesson click or tap on the "take-off Hold" button in the Participants option.

10. Stop file transfer within participants. During meetings, it is possible for participants to share files among each other. This can lead to distraction among participants and sending off offensive images or videos to unassuming participants.

To disable this feature, go to settings, scroll to Personal, then click or tap on the settings option. When it opens, click on the In-Meeting (Basic) option, scroll until you get to File Transfer, then disable it.

You should know that this feature is only available on the web app and not on the desktop app. if you are subscribed to a paid Zoom plan, you have more options. You can disable file transfer for a specific meeting or a particular group depending on your choice.

11. Disable private chat. There's a Zoom in-meeting chat option among participants and between participants and the host. As a host, you can restrict participants from charting privately during the meeting to avoid distractions.

Recording a Meeting

You might want to record your meeting as a way of making sure that people can relate to previous meetings and get the most they can get out of meetings. There is the local recording option available to free and paid subscribers. It allows you to record your meeting and have it stored locally on your computer or device.

After recording and saving it on your local computer or device, the file can then be uploaded to Cloud services like Google Drive or Dropbox for more secure storage. Understand that local recording is not available on Android or iOS. If you are using a mobile device, then the option available is to subscribe to either Zoom Pro Or Enterprise for Cloud recording.

Also, local recording does not allow for timestamp display in recording, audio transcription, and recording of active speakers. If you are looking for these features, then Cloud recording is your best choice.

To record locally

1. Sign in to the Zoom service via the web portal.
2. Go to menu and click on "Settings"
3. When it opens, go to "Recording"
4. Search for local recording
5. Enable the option
6. A verification dialog may show up. Click "Turn On" to confirm you want local recording.

To record during a meeting

1. Start the meeting as a host

2. Click on the record option
3. If various options are displayed, click on "Record on this Computer"
4. A recording icon will show on the top left corner of your monitor or screen
5. When the meeting has ended, Zoom will convert your recording to a compatible file format that you can access.
6. When that is done, a folder with the recording will open up. A video and audio file usually in MP4 format will be named Zoom_0.mp4. If it is an audio only recording then the file format is usually M4A and it is named audio_only.m4a.

To record and save to Cloud, you need to be a paid subscriber. It is possible to record to Cloud using Android or iOS devices. Cloud recording allows you to use different layouts or backgrounds for recordings.

You can also include the gallery view shared screen and active speaker, unlike the local recording.

Prerequisites for Cloud Recording

1. You must be a licensed user

2. You must have a Zoom Pro, Business, or Enterprise subscriber account
3. You must sign in with the Zoom web service, desktop, or mobile app

Your Cloud storage capacity varies according to your plan; Pro users have 1GB storage per user, same for the Business package, while those subscribing to the Education plan have 0.5gb storage capacity per user. For Zoom rooms, the storage capacity is 1gb per Zoom room.

Zoom usually sends a notification when you have reached 80% of your storage capacity. Also, if the capacity is reached during a meeting, the recording will continue until the end of the meeting. You can purchase additional storage for the following prices:

- $40 per month for 100GB, which comes to $1.5 per GB
- $100 per month for 500GB, this comes to $0.5 per GB
- $500 per month for 3TB, this is at a rate of $0.1 per GB

Now that this is out of the way, here is how to enable Cloud recording

1. Sign in to the Zoom web portal
2. Click on the menu and scroll to settings
3. From the options, go to the recording tab
4. Enable Cloud recording if it is disabled

When this is done, you can add some settings to improve the recording experience.

To start and stop a Cloud recording

1. Start the meeting as the host
2. Go to the record button and click on it
3. Select record to Cloud, and the recording will begin
4. To end or pause the recording, click the stop or pause button.

When the recording has ended, Zoom will send a notification email to the host containing two links. The first link is for the host to manage or edit the recording. The second link is for the meeting participants.

Sharing Your Screen

Sharing your screen makes your classes or meetings engaging and fun when visual illustrations on your computer screen are shared with your students or participants.

You can share these during a meeting:

- Your entire desktop screen or mobile screen
- A particular application
- Whiteboard
- Content from a second camera
- Device audio
- iPhone/iPad screen

As a host, you can disable participants from sharing their screen. For Zoom Basic, screen sharing is the default setting for only host only. While in a webinar, only the host, co-hosts and panelists can share their screen.

On Windows and Mac

1. When signed in, click or tap on the share screen option from the Meeting controls.
2. Select which of the screen sharing options you want to enable
   - Basic: This allows you to share your entire desktop screen, a specific application, whiteboard, or iPhone/iPad screen.
   - Advanced: You can either share a portion of your screen, or share your computer audio through your selected speaker in your audio settings. You can also share content from a secondary camera that is connected to your desktop or device.
   - Files from 3rd party devices like Google Drive or Microsoft OneDrive can also be shared. All

you have to do is grant Zoom access to these files
3. When you are done, click share. Your screen will automatically change to full screen to optimize the shared screen view. To exit the feature, click on the Exit Full Screen at the top right corner of the screen or press the Esc key.

Share Screen Menu

1. Mute/Unmute: This menu allows you to mute or unmute your microphone.
2. Start/Stop Video: This allows you to start or stop your in-meeting video.
3. Participants/Manage Participants: This allows you to view and manage meeting participants.
4. New Share: This setting starts a new screen share. When you click on it, a prompt pops up for you to select the new screen you want to share.
5. Pause Share: This allows you to pause your current shared screen.

6. Annotate/Whiteboard: This displays the annotation and whiteboard tools for drawing and adding text etc.
7. More: contains additional settings like:
   - Chat: This opens the chat window
   - Invite: This invites others to join the meeting
   - Record: This begins either local recording or Cloud recording for the screen sharing
   - Allow/Disable Participants Annotation: This allows or prevents the participants from annotating on your shared screen
   - Show/Hide Names of Annotators: This shows or hide the participants' name when they annotate on your screen. When you enable this feature, the annotator's name will briefly show beside their annotations
   - Live on Workplace by Facebook: This shares your webinar or Meeting on Workplace by Facebook
   - Optimize Share for Full-screen Video Clip: This optimizes for a video clip in full-screen video.

Note: if you do not share a full-screen video, do not enable this feature as it blurs your screen share.

- End Meeting: This option allows you to leave the meeting or end the meeting for everyone.

On Android and iOS

The screen sharing feature for Android is only available for devices running on Android 5.0 and above. Screen sharing instructions for Android and iOS are very similar. Also, some options have been disabled in your account settings. Check the integrations option to enable them.

1. Sign in to the Zoom app on your mobile device
2. Start a meeting, then go to the meeting controls and tap on Share
3. Tap the type of content you want to share

   These are the type of content you can share on Android

- Photos: You can select pictures from photo apps or file manager
- Documents: Select images or PDF files
- Dropbox, Google Drive, or Microsoft OneDrive: Select the file from these file-sharing services. You need to grant Zoom access to your account to do this. You can only share PDFs and images.
- Website URL: Enter a URL to open a browser and share the website.
- Share Whiteboard: Share a whiteboard that has annotation enabled.
4. Scroll and tap Screen
5. Tap Start Now
6. Zoom will continue running in the background while the screen share starts.

7. Tap on Annotate at the bottom of the screen to open the annotation tools
8. To end screen sharing, tap on Stop Share

Using Zoom Whiteboard

The Zoom whiteboard is a Zoom for Touch feature that allows you draw, annotate and even invite others to annotate likewise. After a session, you can save drawn whiteboard images and annotations by sending them as a png file to your email or of other recipients you specify. The whiteboard is available for both free and paid Zoom subscribers

What You Need for Zoom Whiteboard
1. Touch drivers for your computer
2. Zoom Rooms for Conference Room for PC, 4.6,0 (1193.1215) or higher
3. Windows PC version 8.1 or higher
4. Single touch screen monitor or dual monitors

Accessing the Whiteboard Feature
1. From the home page after signing into the Zoom service, click the whiteboard option
2. The whiteboard menu bar is shown at the bottom of the screen

Whiteboard Features

- The save feature allows you to mail the whiteboard either as a png file or a pdf file

- The night mode feature depicted by a moon icon changes the color of the whiteboard from white to black
- The close feature closes the whiteboard and returns to the home screen
- The add page feature allows you to add a new whiteboard page. The number of pages opened is shown at the bottom of the page
- The drawing feature helps you to insert lines, shapes and arrows.
- The text feature allows you to insert text in your annotation.
- Start Meeting: This is to start a meeting and share the whiteboard with meeting participants. When the meeting has started, you will see these three options at the bottom right corner.
    - The Micropgone Icon mutes and unmutes the Zoom room's microphone
    - The More icon (…) grants you access to the integrated controller
    - The End Meeting ends meeting for all participants

To invite participants to a whiteboard class, do the following:

1. Tap the whiteboard icon from the home page
2. Tap on your preferred color then start drawing
3. At the bottom right corner of your screen, there's a "start meeting" icon, tap on it
4. Then tap on ... also at the bottom right corner to show the integrated controller.
5. Go to "meeting Controls" and tap on "Invite"
6. Go to the search bar to look for contacts you wish to invite
7. When you have selected the participants you want to invite, tap on the 'invite" button
8. Everyone invited will see the whiteboard and will be able to annotate on it.

Saving a Whiteboard Session

You can save both the whiteboard session and annotations by send ng an email containing the PDF file or image. If there are several whiteboard pages, each page will be a single image. If you select PDF, then the pages will be in a single PDF file.

To save:

- Start a whiteboard session
- Tap on the save icon depicted with the floppy disk in whiteboard controls
- If the whiteboards are more than one, you have to select the ones you want to send
- Enter the destination email address
- At the bottom right corner, select the preferred whiteboard format. If you have several whiteboard pages you want to send in a single file, select the PDF file format.
- Tap on Send and the whiteboard will be sent to the email address.

Using Zoom Annotation

The annotation feature allows you to make notes on the whiteboard of a shared screen. As a host, you can disable annotation for participants to prevent them from scribbling offensive words or, generally, distracting others.

How to Enable Annotation

For Users

1. Sign in to the Zoom app
2. Navigate to Settings
3. Scroll to and click the Meeting button
4. Scroll to Meeting (Basic)
5. Scroll to annotation and enable it
6. Here, you can also enable the feature that allows only the host access to annotation.

For Account

1. Sign in to the Zoom web portal
2. Go to the Menu bar and click on Account Management
3. From the available option, click on Account Settings
4. Scroll and click on the Meeting icon
5. From the options, click on Meeting (Basic)
6. Verify that annotation is enabled
7. If it is disabled, toggle to turn it on.
8. Also, check the box that allows you to save shared screens that contain annotations

For Group

1. Sign in to the Zoom web portal

2. Go to the Menu bar and click on User Management
3. From the available option, click on Group Management
4. Click on the group name, then scroll and click on Settings
5. Scroll and click on the Meeting icon
6. From the options, click on Meeting (Basic)
7. Verify that annotation is enabled
8. If it is disabled, toggle to turn it on. The annotation option may be grayed out in the account settings. You need to enable it there.
9. Also, check the box that allows you to save shared screens that contain annotations

Annotation Tools

After you have shared your screen or enabled the whiteboard feature, annotation tools will be displayed during a meeting. Here are some of the annotation tools:

- Select: This only shows when you have shared your screen or activated the whiteboard. This

tool allows you to move, change the size or select several shapes at the same time
- Draw: This tool is used to insert arrows, lines and shapes. It can also be used for highlighting.
- Eraser: This allows you to clean errors in your annotation
- Undo: To remove your last annotation
- Redo: To restore the last annotation you removed
- PNG: This allows you to save your annotation as a PNG file
- Text: This allows you to insert text
- Stamp: This is to insert icons like a dot, star, or checkmark
- Format: This tool changes the appearance of the annotation tools such as the color, the width of your lines and fonts.
- Clear: The clear tool allows you to delete all annotations at once. If you want to delete only one annotation, use the delete tool or eraser tool.

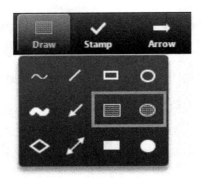

Setting Up Breakout Rooms

The breakout room feature allows your to split your Zoom meeting or class into a smaller and more manageable size. Depending on the size, you can split your room into about fifty separate sessions.

As the host, you can split participants automatically or manually and you can switch among sessions at your choosing.

How to Enable Breakout Rooms
1. Sign in to the Zoom web portal
2. Go to settings from the homepage
3. Scroll to and click on meeting
4. From the options displayed, go to breakout room and click it to ensure it is enabled

5. A dialog box may pop up asking to verify your decision, confirm you want to enable breakout room
6. Once you have enabled breakout rooms, you, as the meeting host, can then manage the breakout rooms as you see fit.

Only the host can assign participants to breakout rooms. Co-hosts can move from different breakout rooms only if they join the room assigned to them by the host. If the Cloud-recording feature is utilized, only the room with the host will be recorded. If it is locally recorded, only the room where the participant is will be recorded. However, multiple participants can record locally.

Depending on the number of breakout rooms, the numbers of participants assigned are given below:

- 20 breakout rooms will hold 500 participants per room
- 30 breakout rooms will hold 400 participants per room
- 50 breakout rooms will hold 200 participants per room

How to Create Breakout Rooms
1. Sign in to the Zoom platform and start a meeting.
2. Click on "breakout rooms."
3. Select the number of rooms you want to create and how you want to assign participants to these rooms
4. You can either select participants automatically or pick them manually one by one
5. When you are done, click on Create Breakout Rooms
6. Your breakout rooms will then be created, but they do not start immediately.

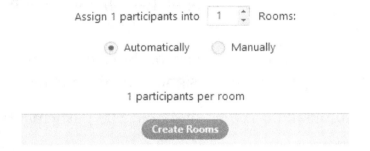

Managing Breakout Rooms
1. After the rooms are created, click on options to see other breakout room options

2. Check the options you want for your rooms. You can move the participants to the room automatically by checking this option. If the option is not checked, then you have to move participants to your desired room manually.
3. If you check the Allow participants to return to the main room at any time option, the participants will be able to move from their breakout rooms to the main room at any time from their end. If the option is unchecked or disabled, then they have to wait in the rooms until the hosts end the breakout rooms
4. Another option is the "Breakout Rooms ends automatically after x minutes." This allows the breakout rooms to end after a predetermined time.
5. "The notify me when time is up" option, when checked, allows the host to be alerted when the time is up
6. The countdown options show the participants how much time they have left before the breakout

room session ends before they are moved to the main room.

After assigning participants to rooms, it is also possible to reassign them if you do not think the arrangement is okay. Participants not assigned to a breakout room will be in the breakout room until the session starts.

Here are some things you can do with your breakout rooms:

1. Move to (participants): You can move a participant to a room you have selected
2. Exchange (participant): You can switch participants from different rooms with each other
3. Delete room: You can delete a selected breakout room
4. Recreate all room: With this, you can delete all previous breakout rooms and start new ones
5. Add room: This allows you to put up additional breakout rooms
6. Open all rooms: This moves all participants to their rooms and starts the breakout sessions. The host will remain in the main room until deciding to join one of the rooms manually.

Participants that have not joined a breakout room will be identified by (not Joined) after their names. You can manually assign them to one, leave them in the main room, or automatically assign them.

When you click on the join icon, you join the breakout room. The leave icon allows you to leave the breakout room and return to the main room. The close all rooms icon closes the entire breakout room after a 60 seconds countdown. All the participants are then returned to the main room.

Breakout room participants can ask for help from the host. They do this by clicking on the Ask For Help button. The host will receive a notification and will then join the breakout room where the request came from.

To broadcast a message to all breakout rooms:

1. Go to the menu option and click on breakout rooms
2. Click on broadcast a message to all
3. Type in your message and click broadcast. All the participants in the breakout rooms will receive your message. Note that whatever you type here

will be seen in all breakout rooms, so you need to be careful with the information you pass across.

Setting Up and Using Polling For Student/Participants

This feature is important for getting feedback from participants. It allows you to structure single or multiple-choice polling questions form participants to answer at your meetings. You can ask the questions during the meeting and download the answers to the polls after the meeting.

You can even make the participants answer the poll anonymously if you think it will help them give you honest answers. For example, if you need feedback on the teaching methods, meeting duration, or even on the topic taught, you can design a questionnaire that reflects these and their feedback will give you an idea of how well or badly you are doing.

As the host, you need Zoom app's desktop version to manage a poll; participants can use any version of the app and any device. Also, only the original host can add or edit polls during a meeting.

If the host or co-host privileges are assigned to another person, they can only launch the original host polls.

To enable the polling feature:

1. Go to settings after signing into the Zoom service
2. Click on Meeting
3. From there, click on Polling
4. If it is disabled, enable it. It possible that a dialog box notification will pop up asking you to verify your decision. Confirm you want to enable the polling feature.

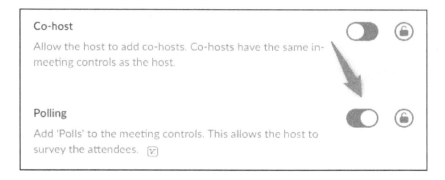

To create a poll:

1. Click on the Meeting icon

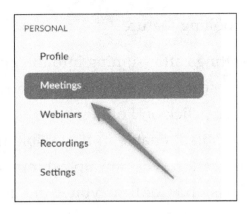

2. Click on scheduled meeting. If there's no meeting scheduled, schedule one

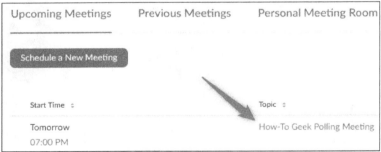

3. Click on meeting management
4. From the options, scroll to the poll option
5. Click add to start a new poll

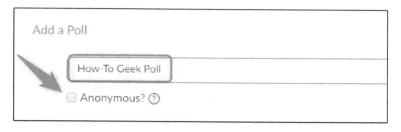

6. Enter your poll title and your first question
7. If you want, check the box to allow participants to answer the questions anonymously, it is not mandatory, though.
8. Select whether you want the participants to choose only one answer or multiple answers
9. Type in the answers to the question and click on save

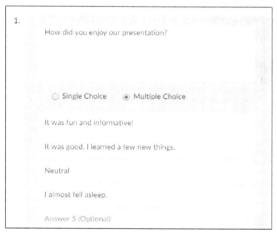

10. To add another question, click on add a question
11. For as many questions as you want to add, just keep clicking on add a question

Note that the maximum poll questions you can ask per meeting is 25.

To launch a poll:

1. Start the meeting with the poll feature enabled
2. Go to the menu bar and click on polls
3. Select the poll you want to launch
4. Click on launch poll

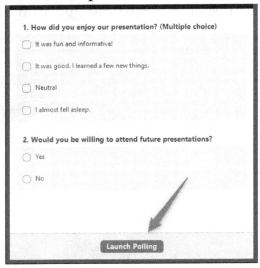

5. All participants will receive a prompt to answer the poll questions. The host will see the results of the poll live.
6. When you want to end the poll, click on end poll
7. If you wish to share the poll results with the students or participants, click on the share results button. The participants will see the results of the polling.

To download a report of a polling result:

1. Sign in to the Zoom web portal
2. Go to account management
3. Then scroll and click on reports
4. Click on usage reports
5. Then click on meeting
6. A list of both previous and upcoming meetings will be shown
7. Scroll and click report type
8. Select poll report
9. Enter the meeting ID of the report you want to download
10. Click on generate

11. You will be directed to a report queue where you can download the poll report as a CSV file

Please note, for meetings where only registered users are allowed, and the poll was not anonymous, the poll participants' names and email addresses will be shown. If the participants were not all registered users, only the profile names of verified users participated in the poll will be shown.

In the case where the polling is anonymous, no names or email addresses will be shown. The word "anonymous" will replace both the name and email addresses of the poll participants.

Setting Up and Using Quizzes For Students/ Participants

Quizzes are essential to know how well the students or participants understand the topic or subject taught in the meeting. It is quite easy to set quizzes for participants or students using the poll feature.

How to Set Up a Quiz

1. Click on the meeting icon
2. Click on scheduled meeting, if there's no meeting scheduled, schedule one

3. Click on meeting management
4. From the options scroll to the poll option
5. Click add to start a the quiz
6. Enter your quiz title and your first question
7. Select whether you want the participants to choose only one answer or multiple answers
8. Type in the answers to the question and click on save
9. To add another question, click on add a question

For as many questions as you want to add, just keep clicking on add a question. Remember that you can only ask a maximum of 25 questions.

Now that the quiz questions are ready, the next thing is to present the quiz during the Zoom class.

1. Start the meeting with the poll feature enabled
2. Go to the menu bar and click on polls
3. Select and launch the quiz
4. All participants will receive a prompt to answer the poll questions. The host will see the results of the poll live.

5. Set a time deadline so the quiz will end automatically when the time is up
6. If you wish to share the quiz results with the students or participants, click on the share results button. The participants will see the results of the quiz.

For a detailed step on how to do this, simply refer to the section to set up and use polling above.

A Short message from the Author:

Hey, I hope you are enjoying the book? I would love to hear your thoughts!

Many readers do not know how hard reviews are to come by and how much they help an author.

I would be incredibly grateful if you could take just 60 seconds to write a short review on Amazon, even if it is a few sentences!

>> Click here to leave a quick review

Thanks for the time taken to share your thoughts!

Using Zoom Chat Tool

The chat tool is available for everyone in a Zoom class or meeting. With this feature, direct messages can be sent to the meeting room, chat messages' settings can be adjusted to allow everyone in the meeting to view the messages, or just allow the meeting host to view the messages.

With Your PC

1. Log in to your Zoom account and ensure you are in a meeting. The chat feature cannot be used if you are not in an ongoing meeting.
2. At the Meeting controls bar located below the meeting window, click the "chat" button. A chat window will be opened at the right-hand side of the screen.
3. In the message box displayed, type a message and click "send" to deliver the message to all participants in the meeting room.
4. To send the message to a specific person or group of persons, click the drop-down arrow beside the message box to display those you can select to send the message to.

With Your Mobile Device

1. Ensure the Zoom app is installed on your device.

2. Open the app and enter a meeting.

3. Tap on "participant" to display the list of activities you can carry out.

4. Tap on "chat" from the list dispayed. From here, craft the message you want to send to those in the meeting or follow through with the procedure discussed above to send the message to a selected few.

Activating Verbal and Non-Verbal Feedback

The verbal and non-verbal feedback feature allows participants to send feedbacks that the host can see through either verbal (words or sound) or non-verbal (emoticons or icons) means. When a participant uses a non-verbal feedback icon, everyone (both the host and other participants) will see it.

To enable non-verbal feedback

1. Sign in to the Zoom web portal
2. Go to Settings

3. Scroll to meeting and click on it
4. Scroll and click on In Meeting (Basic)
5. Toggle the button to enable Non-verbal feedback.

Nonverbal feedback
Participants in a meeting can provide nonverbal feedback and express opinions by clicking on icons in the Participants panel.

You can also make this setting mandatory for all participants by clicking on the lock (key) icon

Nonverbal feedback
Participants in a meeting can provide nonverbal feedback and express opinions by clicking on icons in the Participants panel.

Providing nonverbal feedback (as a meeting participant)

1. Join a Zoom meeting as a participant,
2. Scroll and click on the Participants button

3. Click on the desired icon to provide feedback to the host
4. To remove the icon, click on it again

You can only use one icon at a time. Here are some of the nonverbal feedback icons

- No
- Yes
- Go faster
- Go slower

Clicking on the more button to view additional icons as shown below:

- Disagree
- Agree
- Need a break
- Clap
- Away

When you click on an icon, it will appear next to your name so that the host knows the participant sending the feedback.

To manage nonverbal feedback as a host

1. Click on the Participant button
2. Check the participants' list and their nonverbal feedback
3. The participants' feedback icon will show beside their name
4. If participants raise their hand, you can lower them by hovering over the name and clicking on lower hand

5. To clear all nonverbal feedback, click on Clear all

To receive verbal feedback in Zoom

1. When setting up the meeting, there was the option to mute or unmute the participants' mics.
2. To receive verbal feedback, unmute the participants' microphones, and tell them to do the same from their ends.
3. If anyone wishes to speak, the "raise hand" feature should be used first to permit him. This will help maintain decorum and prevent unnecessary noise in the meeting room.

4. Once permission is granted, he can then speak into his microphone.

Transcribing Zoom Recording

For Zoom Cloud recordings, there is an audio transcript feature that allows you to automatically transcribe your meeting or webinar previously recorded to the Cloud. After processing the transcript, a separate .vtt text file is created in the list of recorded meetings. The recording is divided into parts, with each part having its own timestamp. There is also the option of using the transcript with the recorded video.

If you think the transcribed text is not accurate enough, you can edit it to improve accuracy. It is important to note here that Zoom's transcription service only covers the English language. To have access to this feature, you need to have an active Education, Business or Enterprise license.

You also have to enable Cloud recording and be an account owner or have admin privileges. You might have met all these conditions in some cases and might not have access to the feature. If that is the case, contact the Zoom support.

To enable audio transcript:

1. Sign In to the Zoom web portal
2. Enable Cloud recording
3. Go to advanced Cloud recording settings
4. Scroll to audio transcript and click on the checkbox
5. Save to confirm your selection

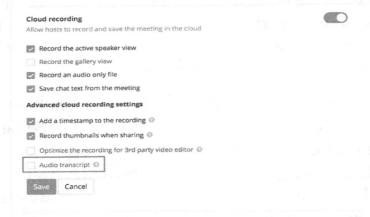

To generate an audio transcript

1. Start a meeting or webinar
2. Record the Meeting to the Cloud

3. At the end of the meeting, you will be sent an email informing you of the availability of the Cloud recording
4. A few moments later, you will receive another mail informing you of the availability of the recorded transcript
5. Both mails will have links to the recording and transcript respectively.

Editing and viewing of transcripts

1. Click the link in the mail or sign in to the Zoom web portal and navigate to "my recordings"
2. Click on the recorded meeting
3. Click on the "audio only" file or the "recording file" as the recording is playing the transcript
4. Scroll to audio transcript on the right hand of the screen
5. Click the pencil icon on the sentence or phrase you want to edit
6. After editing, click the save button

The transcript is automatically embedded in the audio or video file, hidden by default. To see the transcript, go to my recordings and open the file.

Ending a Meeting

Ending a Zoom meeting brings the room's activities to a stop, automatically ejecting every meeting's attendee. Below are the steps to end a meeting.

1. Sign in to the Zoom app and start up your meeting as described earlier.
2. When you are through with your meeting, mouse over to the host controls at the bottom and click on "end meeting for all."
3. This will shut down the meeting activities with all participants ejected. The process is the same, no matter the device and platform used.

Scheduling and Using Zoom Webinar

Zoom allows you to schedule recurring webinars at prearranged dates and times. You can schedule them daily, weekly, fortnightly, or monthly depending on what you want. To schedule a webinar, you need to subscribe to a webinar add-on.

The webinar add-on has different categories 100, 500, 3,000, 5,000 or 10,000. To schedule a webinar you need to:

1. Sign in to the Zoom service
2. Search for and click on webinar
3. At the top of the new page, click on schedule a webinar
4. When the new page displays, enter the topic of the webinar
5. It's not compulsory, but you can enter a webinar description
6. Choose the date, duration, time and time zone of the webinar
7. If you want the webinar to be recurring, enable the recurring webinar option.
8. Depending on the recurring webinars' interval, pick an option of whether, daily weekly, or monthly.
9. Specify if registration for the webinar is mandatory. Also, specify if the participant needs to register only once for the recurring webinars or register per webinar

10. Specify the webinar option
11. Click on the schedule button

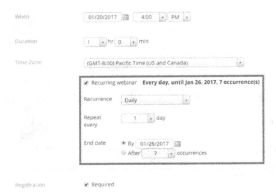

Go to the manage webinar page to see the webinar topic, date and schedule of reoccurrence. To invite participants to your webinar, go to "webinar," and then click on "scheduled webinar." Scroll through the options, and click on "invite attendees."

In addition, enabling the webinar Q & A feature allows for interaction between you and the webinar participants. Like the polling feature, you can make the question and answer segment anonymous.

To activate this feature, go to webinar options, check the Q&A feature, then click schedule. Some webinar features you should consider are:

1. Automatic record of webinar

2. Make webinar available on demand
3. Turn host's and panelists' video on or off
4. Ask password when signing in
5. Save webinar as template to serve as a guide for scheduling future webinars.

Here are some tips for a successful webinar:

1. Publicize your webinar on social media platforms, emails and other platforms like Slack
2. Do a rehearsal
3. Do a proper background design that will appeal and captivate participants
4. Make sure your participants are engaged
5. Ensure you follow up on your participants; this is why it is important to make sure they register so you can have their email address and can send them emails later on

Best Practices While in Zoom Class or Meeting

Some of the best practices to observe in a Zoom class are:

1. Do a dry or practice run before actual classes to get everything in order like audio, camera angle, lighting, internet connection and many others.
2. Look your best
3. Make sure everyone is focused on the topic at hand
4. Secure your Zoom meetings
5. Turn on automatic transcribing
6. Review all controls, enable and disable as you deem fit,
7. Consider making slides beforehand on Canvas for you to teach with if the screen-sharing feature does not work.
8. Anticipate that not all participants will have sound connections, so not everything you say will be heard.
9. Time your class beforehand, especially if you are on the basic plan. This will enable you to stay within the 40-minute time limit.
10. At the start of the meeting, display the meeting or class outline to prepare the class's mind on the meeting or class's objectives and expectations.

11. Use the poll feature to receive participant' responses – to determine if the lesson outline or class is understood.
12. In the case of a large class, use breakout rooms to divide the class into manageable size and organize breakout rooms or groups to encourage participants to work together.
13. Ensure the chat feature is enabled to answer questions asked during the meeting or class. Make sure chat between participants are disabled to discourage distraction

Chapter 4

Troubleshooting Most Common Zoom Problems

As is with every other app, there are issues or problems commonly associated with the usage of Zoom. Below are some of the most common problems and their possible solutions.

1. Webcam or audio not working

It is possible that your device camera is not working or not displaying in conjuction with the Zoom app. If this is the case with you, here are some steps you can take to address the problem.

- Before joining a Zoom call, below the prompt for the meeting ID, there are two options: Do not connect to audio and turn off my video. Make sure you do not check any of the options.
- If the device camera is not displaying, ensure that all other programs that require youdevice's camera are closed. Zoom might not have access to your camera if other applications already has access to the camera.

- If your video or audio isn't working, go to video and audio settings to test both of them
- Ensure that you have given Zoom access to your camera and audio
- Uninstall and reinstall the Zoom app

2. Echoes during call and background noises.

It can be distracting if, during a Zoom call, there is an echo or background noise. Some of the reasons for this are:

- Computer and telephone audio might be active simultaneously. If that is the case, ask the person on the other end to switch off one for the other. The person has to hang up on the call or mute the laptop's audio to achieve this.
- Another reason might be that two people on calls might be too close to each other. Ask them to move further apart.
- It is also possible that computer devices with active speakers are in the same room. In this case,

the two people should also increase their distance.

3. Zoom video freezing or lagging

During a video call, the video feed freezes, or the response time may be slower (lagging). There are some reasons for this:

- The internet connection might be poor, move to somewhere with a better connection. Also, check your internet speed. In meetings with a lot of participants internet speed is very important. If it is possible, look for an internet connection with an average download speed of 1Mbps. The connection should also have an upload speed of 800 Kbps (kilobytes per second).
- You can also get better video quality by changing the video settings. Disabling the High Definition (HD) settings and the touch-up feature will reduce your connection's bandwidth amount and reduce the workload on your device's CPU.

4. Screen sharing problems

Screen sharing is an important feature of Zoom calls or meetings. However, there are some reasons why this might not work.

- Simply ensure that your internet connection is strong enough. Sharing your screen takes a lot of bandwidth.

5. Not receiving emails from Zoom

Emails from Zoom are important; you cannot activate your account, for example, if you do not receive an activation mail from Zoom. There are also notification mails that Zoom regularly send to its subscribers. If you do not receive emails from Zoom, here are some of the reasons.

- Your server might have blacklisted the Zoom IP address. If this is the case, get an IT expert to get them off the blacklist.
- Zoom mails might have been sent to your spam folder. Check the folder and move the mail to the inbox.

6. Zoom app crashing

If the Zoom app crashes or closes suddenly on your device;

- Check your Zoom service status to see if there is a regional problem with Zoom in your area. Zoom may be carrying out maintenance of their servers; this will lead to poor service for some time.
- If the problem is not with the servers, then log in to Zoom via Zoom web instead of using the app.

7. Zoom bombing

Zoom bombing has been extensively discussed above. Kindly refer to the section above on preventing Zoom bombing

8. Missing Zoom features

Some features like screen sharing may be missing during a Zoom call. This is usually because you signed into Zoom using a browser instead of the app. While Zoom works well enough with browsers, its functionality is limited compared to the dedicated app.

If you notice this, use the app or download it if you have not done so already.

Chapter 5

Cool Tips and Tricks to Enhance Productivity With Zoom

Here are some cool tips to help you have a great Zoom experience

1. Use of shortcuts: Learning some cool shortcuts would make your Zoom experience to be sealess and fun. Below are really cool shortcuts that you can deploy when using Zoom for Mac, PC and Linux devices.

Shortcuts for Mac

- Command + J (Join meeting or schedule meeting)
- Command + Control + V (Start meeting)
- Command + Control +S (Share screen through direct share)
- Command +W (Prompt to end or leave meeting
- Command + Shift + A (Mute or unmute audio)

- Command + Control + M (Mute audio for everyone except host)
- Space Bar (Push To Talk)
- Control + Control + U (Unmute audio for everyone except host)
- Command + Shift + S (Start/Stop screen share)
- Command + Shift +T (Pause or resume screen share)
- Command + Shift + N (Switch camera)
- Command + Shift +V (Start or stop video)
- Command +T (Screenshot)
- Command + Shift + R (Start local recording)
- Command +Shift +C (Start Cloud recording)
- Command +Shift + P (Pause or resume recording)
- Command + Shift + W (Switch to active speaker view or switch to gallery view)
- Control + N (View next 25 participants in gallery view)
- Control + P (View previous 25 participants in gallery view)

- Command + U (Display or Hide participants panel)
- Command + Shift + H (Show or hide in-meeting chat)
- Command + I (Open invite window)
- Command + W (Close current window)
- Command + L (Switch to landscape or portrait view)
- Command + Shift + F (Enter or exit full screen)
- Command + Option +Ctrl+ H (Show or hide meeting controls)
- Control + T (Switch from one tab to another)
- Command + Shift + M (Switch to minimal window)
- Command + K (Jump to chat with participant)
- Control + Shift + G (Stop remote control)
- Option + Y (Raise or lower hand)
- Control + Shift + R (Gain remote control)

Shortcuts for PC

- F6 (Navigate among Zoom popup windows)

- PageUp (View previous 25 participants in gallery view)
- PageDown (View next 25 participants in gallery view)
- Alt + F1 (Switch to active speaker view in video meeting)
- Control + Alt + Shift (Shift to Zoom meeting controls)
- Alt + F4 (Close the current window)
- Alt + F2 (Switch to gallery view)
- Alt + H (Display or hide in-meeting chat panel)
- Alt +F (Enter or exit full screen)
- Alt + U (Display or hide participants panel)
- Alt + L (Switch to portrait or landscape view)
- Alt +I (Open invite window)
- Control +F (Find or search)
- Control + Alt + Shift + H (Show or hide floating meeting controls)
- Control + Tab (Move to next tab)
- Control + Shift + Tab (Move to the previous tab)
- Alt + V (Start or stop video)

- Alt + M (Mute or unmute audio for everyone except host)
- Alt + A (Mute or unmute audio)
- Alt + S (Launch or stop share screen window)
- Alt + Shift + S (Start or stop new screen share)
- Alt + T (Pause or resume screen share)
- Alt + R (Start or stop local recording)
- Alt + C (Start or stop Cloud recording)
- Alt + P (Pause or resume recording)
- Alt + N (Switch camera)
- Alt + Shift + T (Screenshot)
- Alt + Y (Raise or lower hand)
- Control + W (Close current chat session)
- Alt + Shift + R (Gain remote control)
- Alt + Shift + G (Stop remote control)
- Control + Up (Go to previous chat)
- Control + Down (Go to next chat)
- Control + T (Jump to chat with someone)

Shortcuts for Linux

- Alt + V (Start or stop video)
- Alt + A (Mute or unmute my audio)

- Alt + T (Pause or resume screen sharing)
- Alt + S (Start or stop screen sharing)
- Alt + M (Mute or unmute audio for everyone except host)
- Alt (To toggle the always show meeting controls option)
- Alt +P (Pause or resume recording)
- Alt + U (Show or hide participants panel)
- Alt + N (Switch camera)
- Ctrl + Tab (Switch from one tab to another)
- Alt + R (Start or stop local recording)
- Alt + C (Start or stop Cloud recording)
- Alt + Shift + T (Screenshot)
- Esc (Enter or exit full-screen mode)
- Alt + I (Open invite window)
- Alt + Y (Raise or lower hand)
- Control + W (Close current chat session)
- Alt + Shift + G (Revoke remote control permission)
- Alt + Shift + R (Begin remote control)

2. Virtual background: Since you will be doing video conferencing most of the time, the appearance of the background is important, especially if you are teaching or organizing a meeting from home. You can use various images from those available on the Zoom app or you can import your own image from your local device.

To activate virtual background using the mobile Zoom app:

- Sign in to your Zoom account
- Start or join a meeting
- Click on the ... at the right bottom of your screen to open the More menu
- Tap on Virtual Background
- Select a background from those available
- To select a customized background, click on the + icon. A box will appear with pictures on your device; pick your preferred picture.

To activate virtual background using the desktop Zoom app:

- Sign in to the Zoom app, click on your profile and scroll and click on settings
- Scroll and click on Virtual Background
- Select from the default backgrounds available
- To select a customized background, click on the + icon. A box will appear with pictures on your device; pick your preferred picture.

3. Meeting reminders (mobile): This feature reminds you of upcoming Zoom meetings. With your busy schedule, it is possible to forget an upcoming meeting. With the upcoming reminders feature, yodon't have to worry about forgetting upcoming events. To enable this feature, you have to integrate your calendar with the Zoom app.

To enable calendar integration:

- Sign in to the Zoom web portal
- Click on your profile
- Scroll to the Calendar and Contact Integration
- Scroll and click on Connect to Calendar and Contact Service

- Select a service (Google, Office365, Exchange etc.)
- Change the permissions for service
- Click on the Next button
- Follow the instructions and confirm to grant Zoom access to your calendar and contacts.

Enabling meeting reminders:

- Click on Account Management
- Scroll and click on Account Settings
- Scroll to the Meeting tab and turn on the Upcoming Meeting Reminder

4. Gallery View. There are three different video layouts; active speaker view, gallery view and floating thumbnail window. The active speaker shows you the host or the co-host, while the gallery view shows the participants. If there are 49 people or less, all participants will show on a single screen. As a host, it is better to use the gallery view to see all your participants.

To activate the gallery view:

- Sign in to the Zoom service
- Start a meeting
- Click the view icon at the top right corner
- Select the gallery option

The end... almost!

Hey! We've made it to the final chapter of this book, and I hope you've enjoyed it so far.

If you have not done so yet, I would be incredibly thankful if you could take just a minute to leave a quick review on Amazon

Reviews are not easy to come by, and as an independent author with a little marketing budget, I rely on you, my readers, to leave a short review on Amazon.

Even if it is just a sentence or two!

So if you really enjoyed this book, please...

\>\> Click here to leave a brief review on Amazon.

I truly appreciate your effort to leave your review, as it truly makes a huge difference.

Chapter 6

Zoom Frequently Asked Questions (FAQs)

Here are some of the frequently asked Zoom questions to help you get the hang of what other Zoom users typically ask.

Do I need an account to use Zoom?

As a participant and in most cases, you do not. However, in some cases, the meeting hosts may bar participants without an account from joining the meeting. If you want to host a meeting, however, you need a Zoom account.

How do I purchase a webinar license?

To host a webinar, you need to be a Zoom licensed user and have a webinar license. Visit the billing page of Zoom on https://Zoom.us for more information.

How do I reset my password

Visit https://Zoom.us/forgot_password and follow the steps to reset your password.

Can I record my meeting?

All Zoom meetings can either be recorded locally (on your computer) or to the Cloud depending on your subscription plan or type of device. If you are using an Android or iOS device, then you can only use the Cloud record option. If you are on the Basic plan, then you can only record locally.

Where can I download the latest version of Zoom?

To download the latest version of Zoom visit https://Zoom.us/download or go to your Google and Apple app stores.

How many people can attend my meeting?

Depending on your package, 500 people can attend your meeting at a go. Zoom Basic is free and allows meetings for 100 participants. Zoom Pro costs $149.90 and allows meetings with 100 participants, but unlike Zoom Basic, it offers 1GB Cloud recording, and there's no time limit.

The Zoom Business package costs $199.90 per year per license. You can host up to three hundred participants and it comes with a single sign-on Cloud recording. Zoom Enterprise costs $199.90 and allows you to host

500 participants at once. It also offers unlimited Cloud recording storage.

Are Zoom meetings encrypted?

By default, all content from Zoom in-meetings and in-webinars are encrypted. As a host, you can add another layer of security by enabling end-to-end encryption setting. This setting enforces encryption across all Zoom enabled devices.

How safe is my Zoom recording when saved to Cloud?

As a meeting host, you alone have access to your Cloud recordings. The only way others can access it is to download it or send the recording URL to others. There's also an option where you can allow authenticated users to view the recordings, or you can add a password to the recordings or mandate that users register before viewing the recording.

What is a Zoom meeting ID and where can I find it?

This is a 10 or 11 digit number that is associated with a scheduled or instant meeting. If it is an 11-digit number, then it is used for a recurring, scheduled, or instant meeting while a 10 digit number is the Personal Meeting ID (PMI).

The meeting ID for an instant meeting expires after the meeting ends. That of a non-recurring scheduled meeting expires 30 days after the scheduled meeting date for a non-recorded meeting.

If the scheduled meeting is restarted, the ID is renewed for another 30 days. For recurring meetings, the ID expires if there's no meeting in 365 days.

In the Zoom desktop app, the meeting ID, passcode and join link are found when you click the info icon at the top corner of your screen. For the Zoom mobile app, the ID appears when you click the top of the screen.

Conclusion

Video conferencing is the new norm in today's world. With advancements in technology and the relatively cheap cost of holding virtual meetings and classes via the Zoom service, there is no doubt that this is a platform all teachers, coaches, instructors and business owners should have at their disposal.

Zoom is easy to use, which allows for a seamless and engaging experience with your students, business partners and co-workers alike. It comes with several cool features that have been extensively discussed in this book, such as screen sharing, annotation, polling and whiteboard, to mention but a few. Zoom allows you to record your class sessions locally or to the Cloud for future reference, depending on your subscription package.

The Zoom breakout room feature also allows you to reduce your class size to a manageable number. Likewise, you can set quizzes for students and enable and encourage verbal and nonverbal feedback. There is just a whole lot that the Zoom service can provide.

Zoom is part of the modern world and it has helped shaped how meetings and classes are organized. So then, embrace it wholeheartedly to deliver the best virtual learnings and online meetings you can possibly think of. I hope the pages of this book have been your companion to navigating your way around how to put this great tool to excellent use in your journey toward providing an interactive and fun virtual learning and meeting experience for your students, business partners and co-workers.

I wish you all the best!

CPSIA information can be obtained
at www.ICGtesting.com
Printed in the USA
BVHW041045110321
602311BV00004B/246